SNAP, BUTTON, ZIP
Inventions to Keep Your Clothes On

VICKI COBB
Pictures by Marylin Hafner

HarperTrophy

A Division of HarperCollins*Publishers*

Library of Congress Cataloging-in-Publication Data
Cobb, Vicki.
  Snap, button, zip : inventions to keep your clothes on.

  Summary: Presents simple historical background on the
things that fasten our clothes, such as elastic, zippers,
buttons, and sticky tapes.
    1. Fasteners—Juvenile literature.   2. Children's clothing—
Juvenile literature.   [1. Fasteners.   2. Clothing and dress]
I. Hafner, Marylin, ill.   II. Title.
TT557.C63   1989   646'.19   87-26097
ISBN 0-397-32142-2
ISBN 0-397-32143-0 (lib. bdg.)
ISBN 0-06-446106-8 (pbk.)

First Harper Trophy edition, 1993.

Time to get dressed in the morning.

How fast can you do it?

You pull on your underwear. *Pop!* goes
the elastic at your waist.

You put on your shirt.
It takes two hands to push
each button through its hole.

You pull on your pants. Zip the zipper. Snap the snap.

Pull on your socks.

Push your feet into your shoes.
Press the sticky tapes of your
shoes closed.

5

There! You're ready.

Your clothes are on for the day.

Suppose the elastic and zippers and buttons and sticky tapes suddenly disappeared.

You'd have to wear different kinds of clothes. Maybe clothes that tied on.

Maybe clothes that wrapped you tight.

You could always wear styles from different parts of the world.

It would take a lot longer to get dressed. You'd probably have to use lots of pins and belts, and you'd still have problems.

Many people have thought about how to keep our clothes from falling off. Buttons, elastic, zippers, and sticky tapes that press closed all began as an idea in someone's head. This is how they work.

BUTTONS

A button is almost the same size as the hole or loop it is poked through. Once a button is pushed through a hole and the cloth lies flat, the button does not slip out.

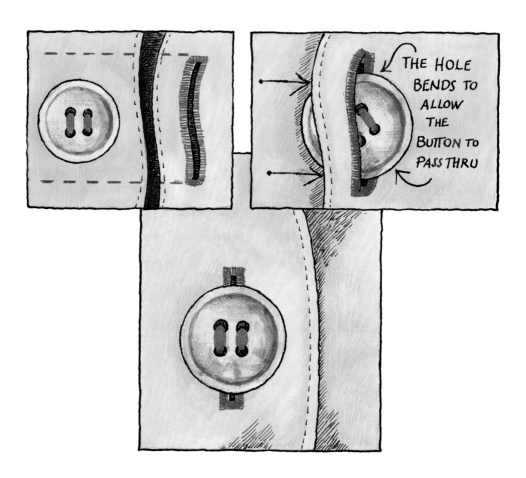

THE HOLE BENDS TO ALLOW THE BUTTON TO PASS THRU

We don't know who invented them, but buttons have been around for thousands of years. The earliest buttons were made of bone or shell.

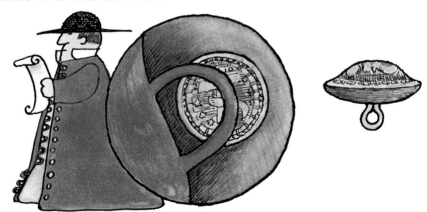

At first buttons were pushed through loops that were attached to clothes.

Then hundreds of years later, someone thought of putting a hole in the piece of clothing. The button could now fasten clothes anywhere there was a buttonhole.

NOW CLOTHES CAN FIT AND REALLY KEEP YOU WARM.

The buttonhole made buttons very popular. People wore buttons of wood and bronze. Rich people wore buttons of gold and silver and precious stones.

SIRE, THE ROYAL BUTTON-VENDOR IS HERE TO SHOW HIS WARES.

Many buttons were only for decoration. But the buttons sewn on men's coat sleeves did have another job to do. They were supposed to stop men from wiping their mouths on their sleeves when they ate.

Buttons were once used to close shoes. Certain shoes were called high-button shoes. People needed a special tool called a buttonhook to close their shoes.

BUTTON HOOK–CIRCA 1900.

A VERY USEFUL TOOL!

Most of the buttons on our clothes today are made of plastic or metal. They are not fancy or expensive. But a lost button is missed!

Some people collect buttons. Perhaps there is a button box in your family. You can have fun playing with a button box and hearing the stories of special buttons and the people who wore them.

ELASTIC

If you want clothes to fit closely, they have to open up so you can get them on. And then they have to somehow get smaller. Elastic does this job to perfection!

Elastic is threads made of rubber, just like the rubber in a rubber band or ball. Rubber snaps back to its original shape no matter how it is pushed or pulled. So you can stretch it or twist it, but when you let go, it's always the same shape.

Where can you find elastic?

The waistband on your underwear is elastic tape.

Your socks have elastic threads in them to hold them up.

Tight-fitting bathing suits have elastic threads knitted into the fabric.

Sometimes you can find elastic at your wrists and ankles.

ZIPPERS

Almost 100 years ago Mr. Whitcomb Judson of Chicago had a friend who suffered from a bad back. His friend couldn't bend over and use two hands to button his shoes. Mr. Judson wanted to invent a fastener that his friend could use with one hand.

Perhaps Mr. Judson got his great idea from watching a bird smoothing its feathers. A feather is made of many strands that hook together. It's easy to pull apart a feather so that the strands separate.

If you push the strands together and pinch your fingers along the separation, you can make the feather whole again. Your sliding fingers make the tiny hooks on one strand hook into its neighbor. A bird hooks a feather together by running its beak down the split.

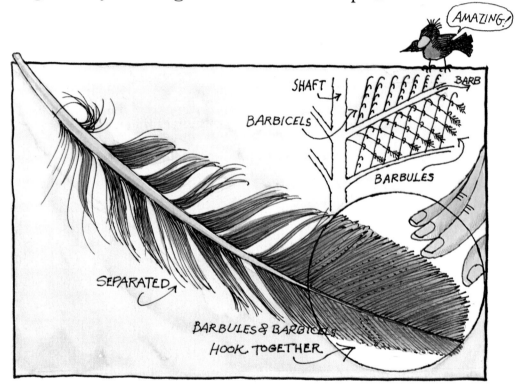

Mr. Judson invented two rows of teeth that hooked together when a slide was moved.

The trouble with his fastener was that it came open unexpectedly.

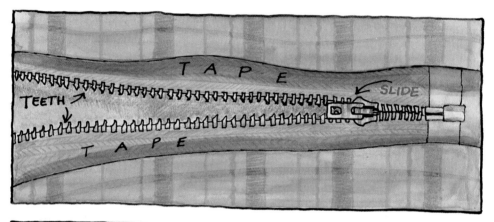

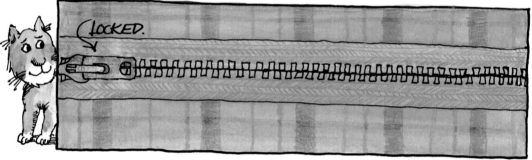

Twenty years later, another inventor came to the rescue. He made a zipper with teeth on two tapes. The ends of one set of teeth are too fat to slip into spaces between the ends of the opposite set. The slide bends the tape so the teeth are spread apart. Now the teeth can slip between each other. After the slide passes, the tape is straight and the teeth firmly grip each other. The zipper is locked closed.

STICKY TAPES

Ever come home from a walk in the country covered with burrs? These pesty burrs stick to anything that brushes by them. There are seeds in a burr. When people and animals pick up burrs, they spread the plant's seeds.

SEEDS SCATTER TO OTHER PLACES AND TAKE ROOT IN THE GROUND. NEW TREES GROW.

One day about fifty years ago, some burrs stuck to George de Mestral, a Swiss engineer. Mr. de Mestral wondered what made the burrs stick.

He put a burr under the microscope and saw thousands of tiny hooks.

Mr. de Mestral invented two kinds of tape. One kind has thousands of tiny hooks, just like the burr. The other is covered with thousands of loops. Press them together and some of the hooks grab some of the loops. The tapes stick together instantly. Pull them apart and the hooks unbend enough to come out of the loops. These tapes can open and close things over and over.

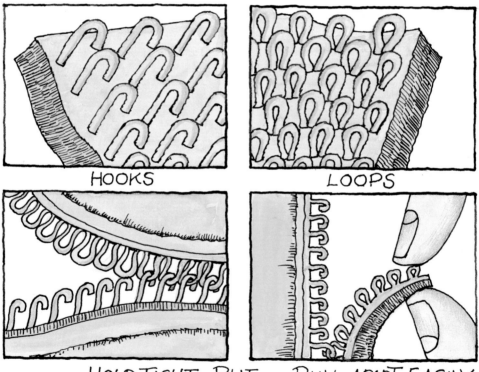

HOOKS

LOOPS

HOLD TIGHT BUT... PULL APART EASILY.

Sticky tapes may be sewn in many places where your clothes close.

They are also used to hold lots of other things in place.

THE QUILT COVER ON YOUR BED

YOUR MOTHER'S WALLET

YOUR SNEAKERS

THE SLIPCOVER ON DAD'S CHAIR

UNCLE IRVING'S CAST

TOOLS IN THE GARAGE

Clothes may also be fastened by snaps, laces, or hooks and eyes. Each kind of fastener does its own job.

COSTUMES OF OLDE ENGLAND

Each helps make clothes comfortable, well fitting, and stylish.

 THE END